Natural Disasters

AVALANCHES AND LANDSLIDES

JANE WALKER

GLOUCESTER PRESS
London · New York · Toronto · Sydney

CONTENTS

INTRODUCTION

A huge mass of snow, rock or mud crashes down from a mountain or hillside at high speed. It can engulf and destroy entire towns and villages in just a few minutes. With a thunderous roar, avalanches and landslides strike quickly and often without warning, and their impact can be catastrophic.

Hundreds of avalanches occur each day on remote snowy mountaintops throughout the world. However, their effect is only really felt in the populated mountain regions. Here, the whirling mass of ice and snow will crush to death or bury alive any person or animal unlucky enough to be caught in its path.

The introduction of improved warning systems can reduce the effects of avalanches and landslides. But we must also try to control the human activities that can trigger these natural disasters.

Although the term "avalanche" can mean the fall of any material – snow, soil, rocks, ice or volcanic lava – it generally refers to a falling mass of snow. Dry-snow avalanches can travel at speeds of up to 360 km/hr. The melting snow of a wet-snow avalanche moves more slowly, at about 35 km/hr.

Wet-snow avalanche
Wet-snow avalanches often begin in rocky areas. Here, the snow melts around rocks that have warmed up in the sunshine. The fresh, wet snow can set hard, like concrete, when the avalanche finally comes to a halt.

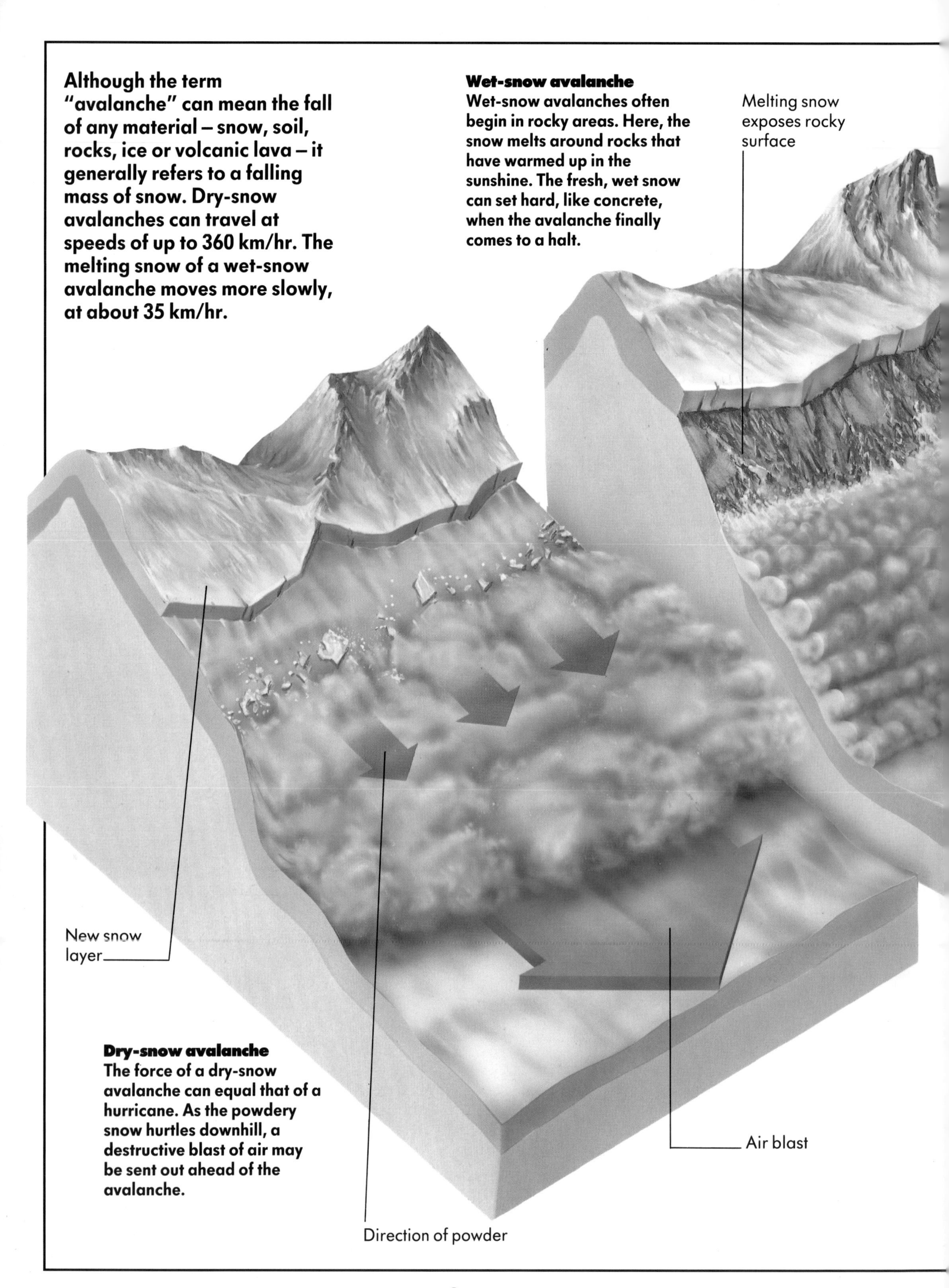

Dry-snow avalanche
The force of a dry-snow avalanche can equal that of a hurricane. As the powdery snow hurtles downhill, a destructive blast of air may be sent out ahead of the avalanche.

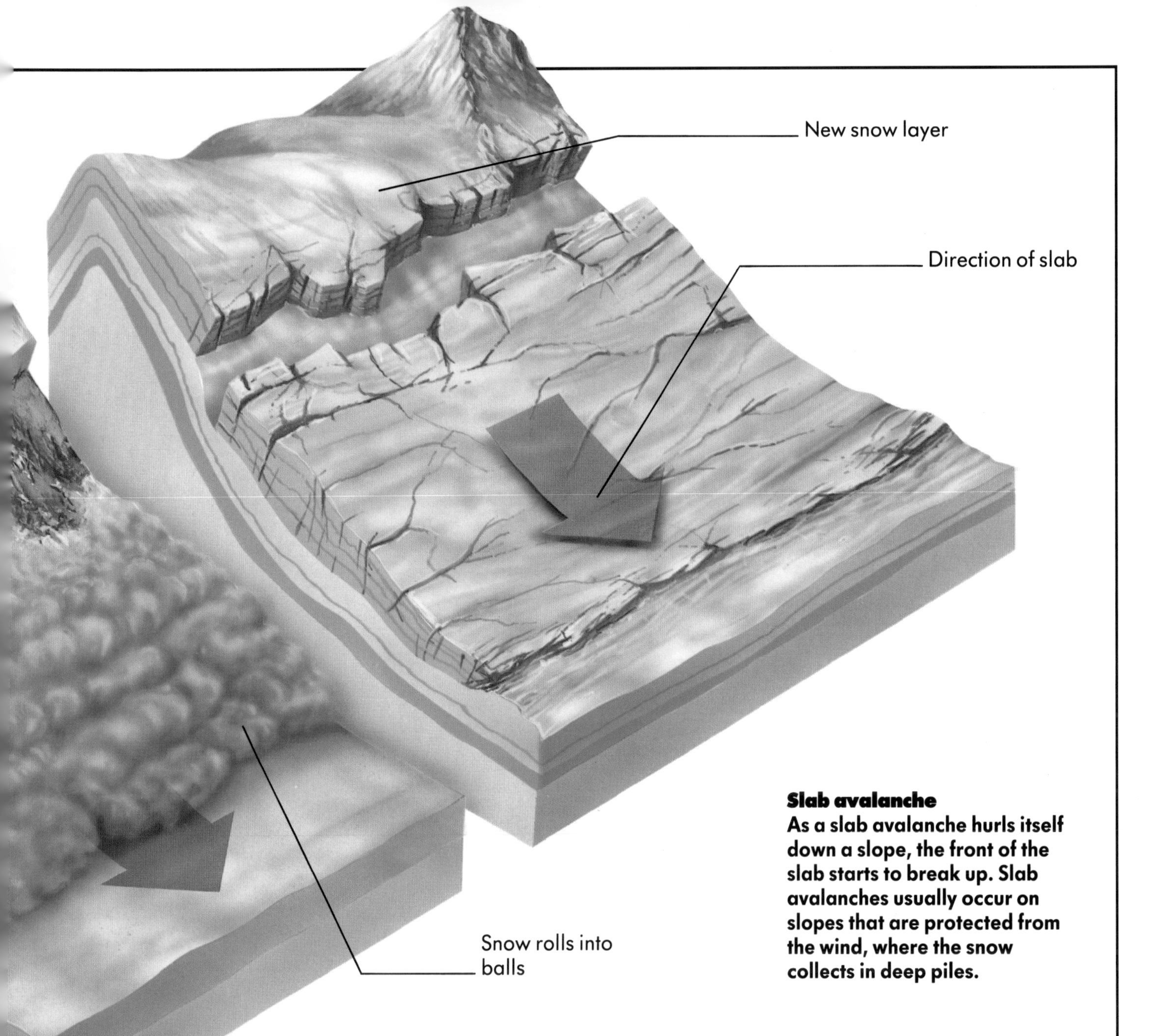

Slab avalanche
As a slab avalanche hurls itself down a slope, the front of the slab starts to break up. Slab avalanches usually occur on slopes that are protected from the wind, where the snow collects in deep piles.

WHAT IS AN AVALANCHE?

An avalanche is a huge mass of ice and snow which breaks away from the side of a mountain and surges downwards at great speed. The greatest avalanches probably occur on the high peaks of the Himalayas. However, those which cause the highest death toll fall in the populated valleys of the Alps.

Scientists have grouped avalanches into three main kinds: wet-snow avalanche, dry-snow avalanche and slab avalanche. Wet-snow avalanches usually occur in the spring, when the loose, melting snow forms into large boulders of snow as it rolls downhill. More deadly are the dry-snow avalanches, which either slide along close to the ground, or lift off the ground completely and swirl through the air, often hundreds of metres high. In a slab avalanche, a huge chunk of solid, sticky snow breaks away from a slope. It slides across a layer of loose snow crystals lying beneath the surface.

AN AVALANCHE BEGINS

Most avalanches begin either during or soon after a snowstorm. As each new layer of snow settles on the ground, it binds itself to the existing layers that are anchored to the mountainside. The additional weight of a heavy snowfall can prevent the snow from gripping onto the layers of snow underneath, and trigger an avalanche. Avalanches also occur during the spring, when melting snow seeps down through the surface. This creates a slippery layer on which the snow can slide.

The steepness of the slope can also affect the speed of the avalanche. Steep, rocky slopes help to anchor the snow. The smooth surface of gentler grass-covered slopes allows the avalanche to reach great speeds.

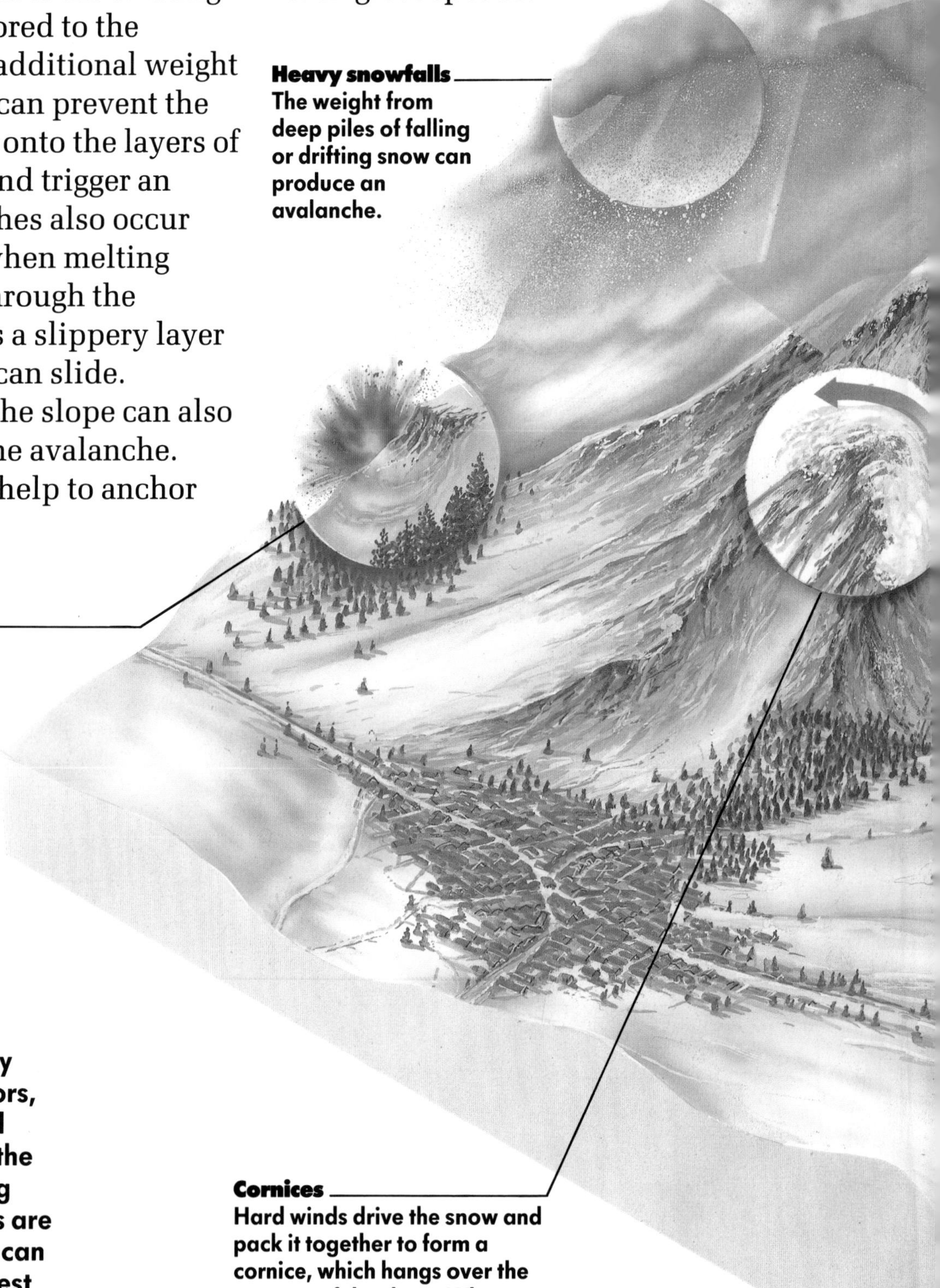

Heavy snowfalls
The weight from deep piles of falling or drifting snow can produce an avalanche.

Explosions
An avalanche can be set in motion by an explosion at a mine or quarry, by the noise of a low-flying plane, or even by a clap of thunder.

Avalanches are not only caused by natural factors, such as the amount and quality of the snow, or the sudden arrival of strong winds. Once conditions are suitable, an avalanche can be started by the slightest movement, such as a falling icicle or twig, or by a small animal crossing a slope.

Cornices
Hard winds drive the snow and pack it together to form a cornice, which hangs over the very top of the slope. When a cornice falls, it can trigger an avalanche.

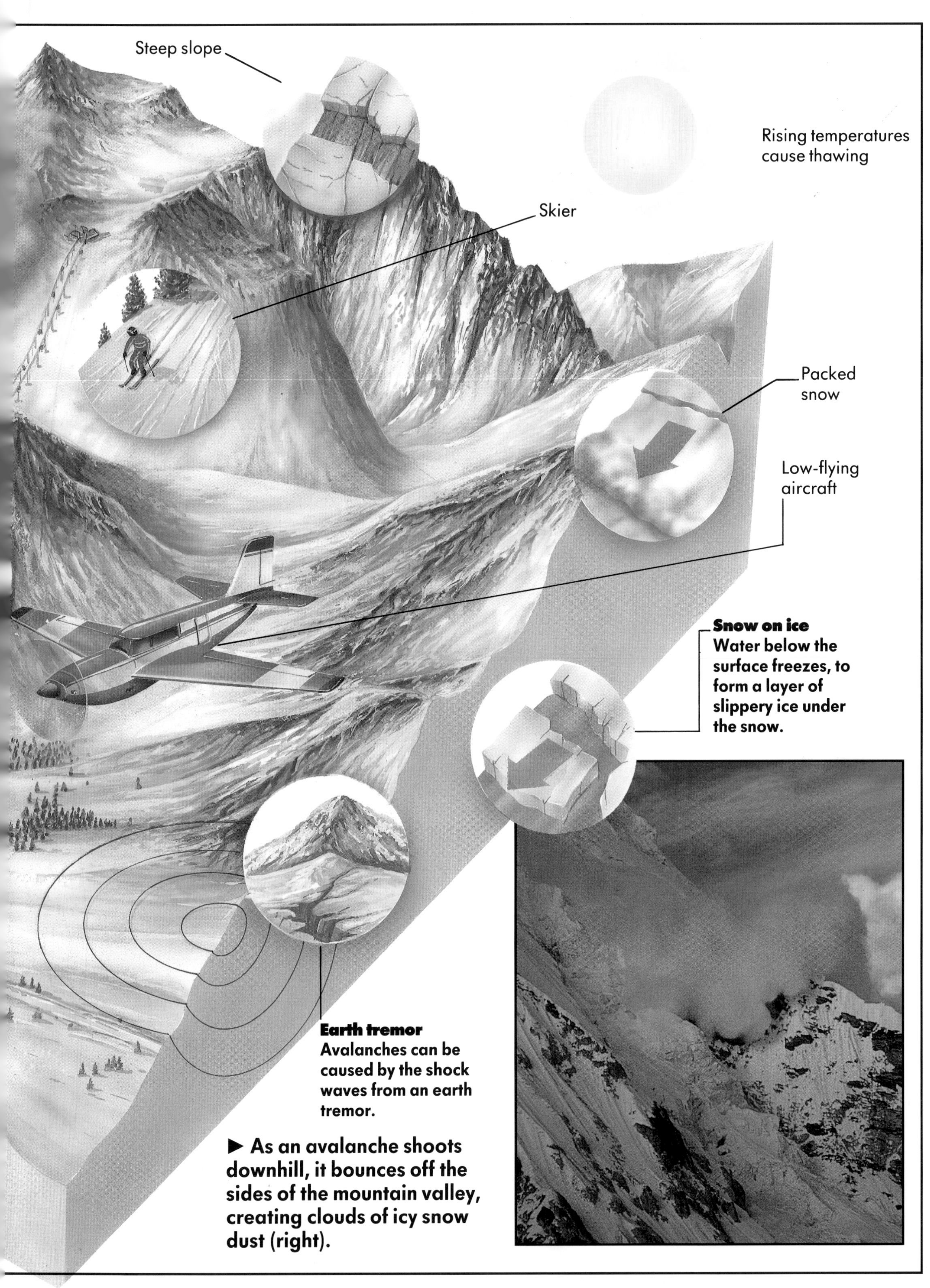

► As an avalanche shoots downhill, it bounces off the sides of the mountain valley, creating clouds of icy snow dust (right).

THE WHITE DEATH

When an avalanche strikes, more than 1 million cubic metres of snow and ice blast down a mountainside. As the ice mass falls, it collects large amounts of debris such as rocks and tree stumps on its way. In a populated mountain area, whole villages are crushed, hundreds of people and animals are buried alive, power and water supplies are cut off, and roads and railway lines disappear – all in a matter of minutes.

The world's greatest single avalanche disaster occurred in 1962 in Peru. More than 3,500 people died and eight villages and towns were destroyed in just 7 minutes. When the avalanche finally came to rest, after a journey of 15 kilometres, the pile of snow and ice was over 18 metres deep.

Trapped under the hard, packed snow, an avalanche victim can barely move and is unlikely to survive for more than a couple of hours. Victims die from the cold, from a lack of oxygen, or from the injuries that occurred when they were first struck. Only about 5 per cent of all avalanche victims are rescued alive.

◄ In the mountain village shown left, a path has been cut through the wall of snow and debris caused by a recent avalanche.

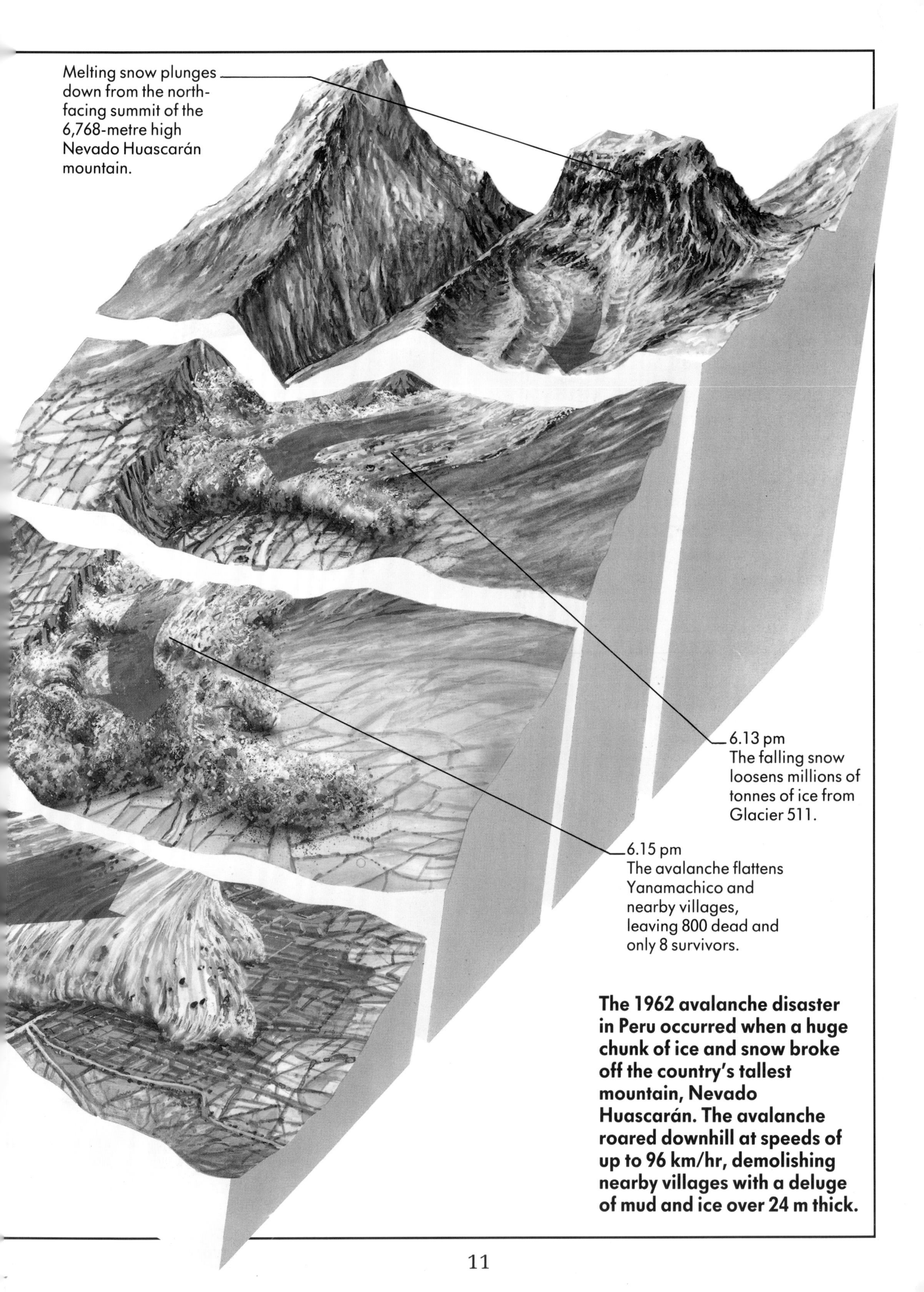

The 1962 avalanche disaster in Peru occurred when a huge chunk of ice and snow broke off the country's tallest mountain, Nevado Huascarán. The avalanche roared downhill at speeds of up to 96 km/hr, demolishing nearby villages with a deluge of mud and ice over 24 m thick.

THE ALPS

Each year, tens of thousands of avalanches occur in the European Alps, the world's most densely populated mountain region. The inhabitants of the numerous Alpine towns and villages live under the constant threat of avalanches, which is made worse by certain natural conditions that prevail there.

The warm dry Föhn wind which blows through the deep Alpine valleys raises temperatures suddenly, causing the snows to melt rapidly. The smooth surface of the Alpine meadows lubricates the path of the avalanches, which travel at terrifying speeds. In 1898, an avalanche at Glärnisch in Switzerland raced downhill at around 400 km/hr.

During the 1950-51 "Winter of Terror" in Switzerland, over 1,100 avalanches killed a total of 98 people.

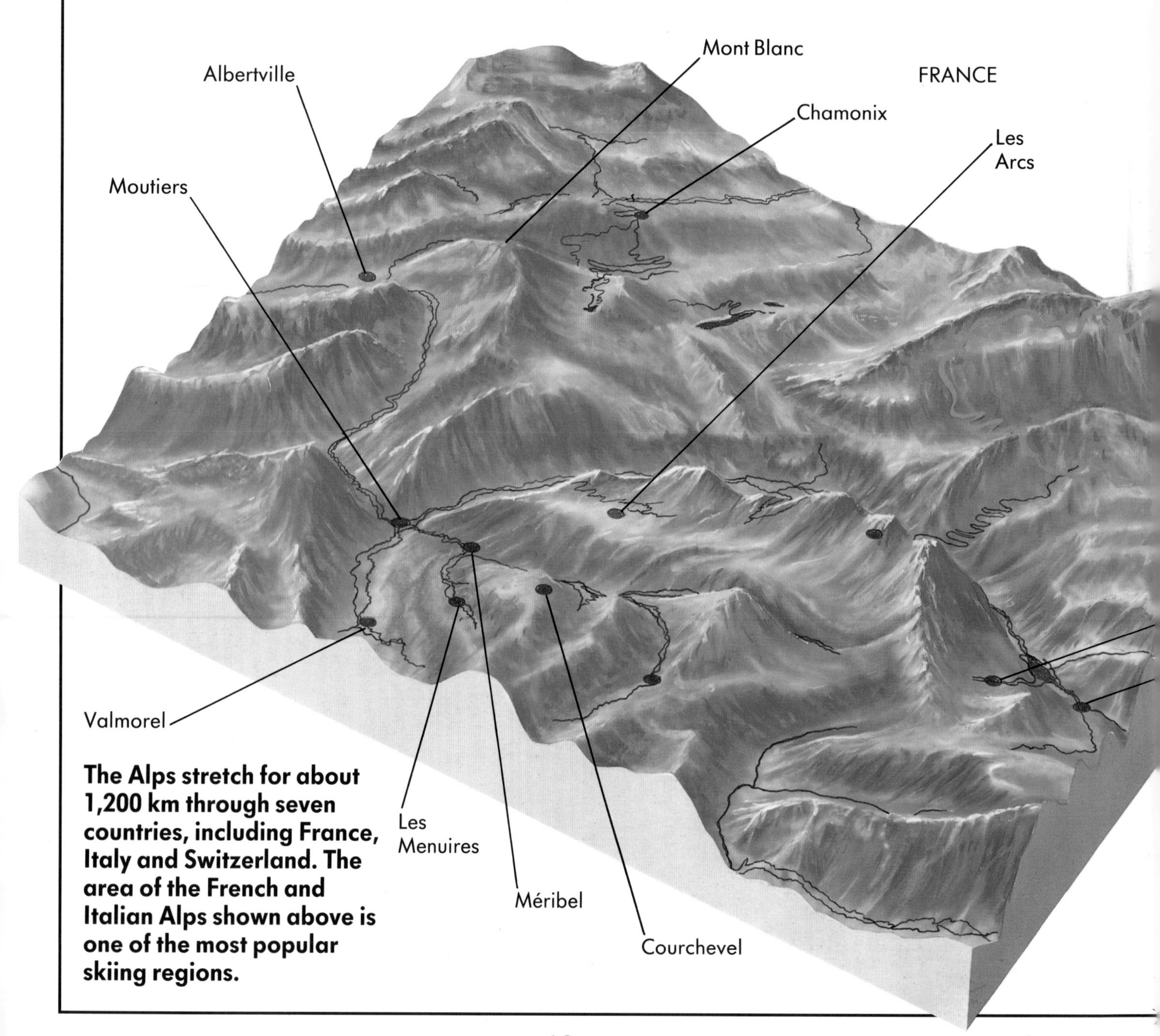

The Alps stretch for about 1,200 km through seven countries, including France, Italy and Switzerland. The area of the French and Italian Alps shown above is one of the most popular skiing regions.

◀ Avalanche warning signs, such as the one shown left, inform skiers if there is a danger of avalanches. The ski trails, called pistes, that are at risk are then closed.

Experienced skiers who choose to ski "off piste" on unmarked trails run the risk of setting off an avalanche. The weight of a single skier on a slope can open up a crack in the hard crust on the snow's surface. This releases a huge slab which breaks away, often at a point above the skier, offering little chance of escape.

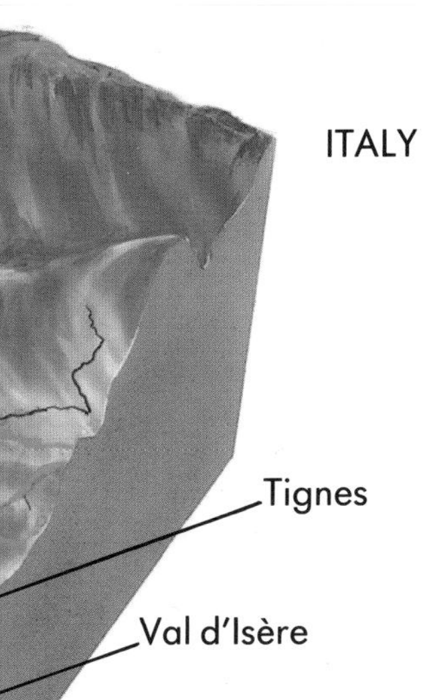

▶ More and more people visit the Alps each year. New hotels, ski lifts, electricity lines and roads are being built to meet the increased demand for facilities. Avalanches in these developed areas are now more likely to cause widespread damage and many deaths.

WHAT ARE LANDSLIDES?

A landslide is the name given to the downhill movement of large amounts of soil, mud, rocks and other debris. There are three different kinds of landslide: falls, slides and flows.

In a rockfall, huge boulders crash down a steep slope, often breaking up into smaller pieces on reaching the ground. When rockslides occur, the chunks of rocky material slide over the ground as quickly as flowing water. Debris slides, which consist of thin layers of loose soil and smaller rocks, travel in the same way.

Mudflows and earthflows are wet landslides. A mass of mud and water flows down from the upper slopes, picking up debris lying in its path.

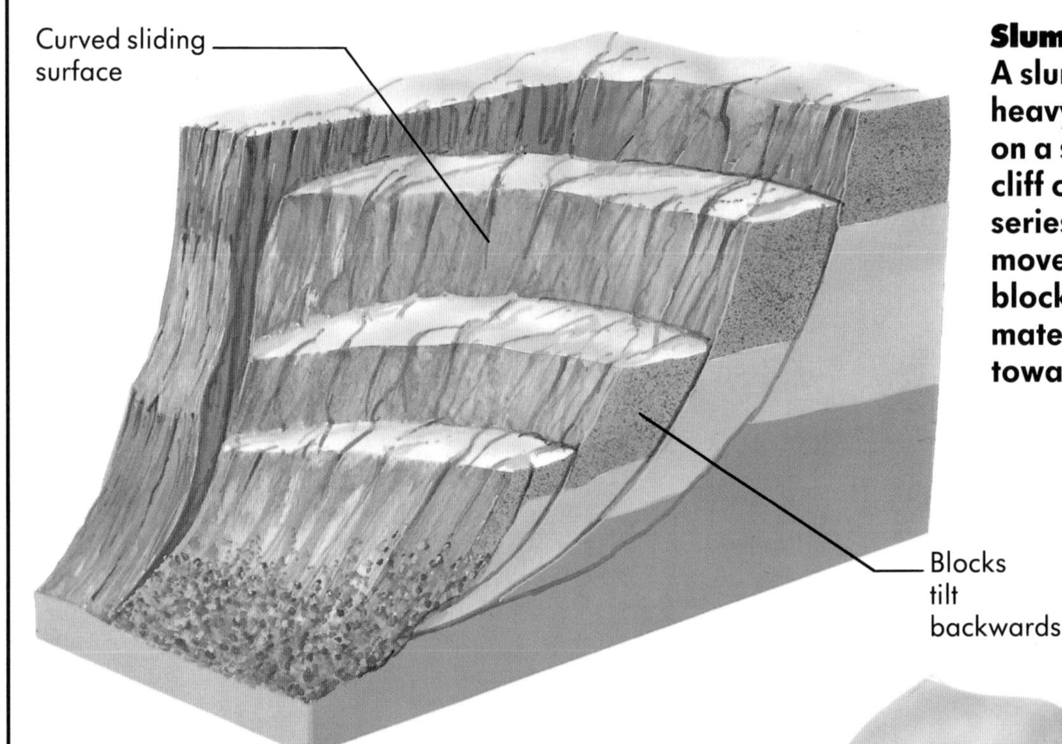

Slump
A slump occurs when heavy rock and soil on a steep slope or cliff collapse in a series of curving movements. The blocks of slumped material tilt back towards the slope.

Rockslide
Rockslides involve the downward movement of rock debris or large blocks of rock. They often occur after heavy rains and in areas where large numbers of trees have been removed from the slopes.

Large blocks become detached from slope

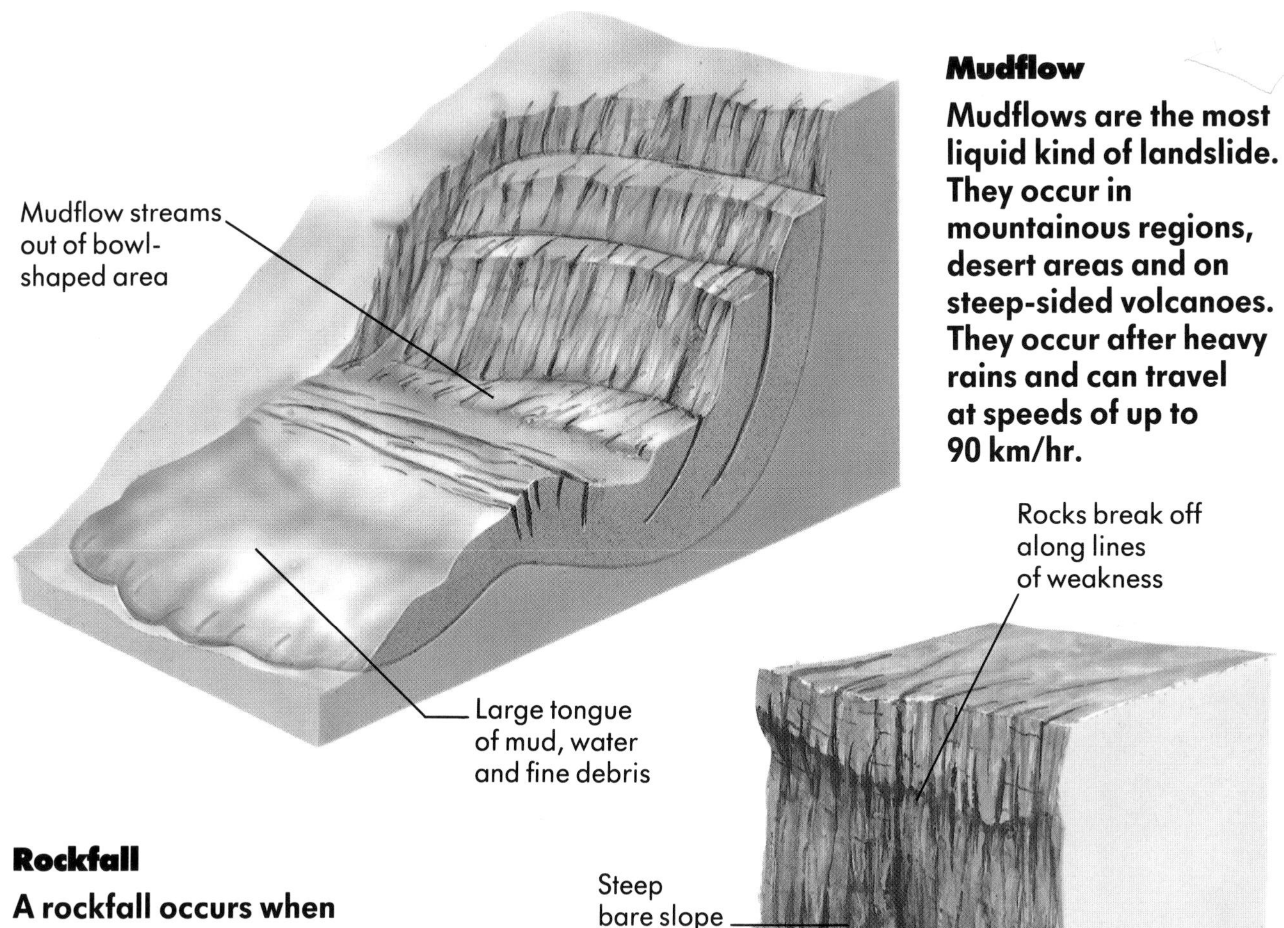

Mudflow

Mudflows are the most liquid kind of landslide. They occur in mountainous regions, desert areas and on steep-sided volcanoes. They occur after heavy rains and can travel at speeds of up to 90 km/hr.

Rockfall

A rockfall occurs when rocks descend at high speed, often falling freely through the air, down a steep slope or cliff face. The rocks may have been loosened by frost or ice. On reaching the lower slopes, a rockfall may break up into a moving torrent of debris.

Flooding
Flooding is a common after-effect of landslides. The debris from a landslide can fill a riverbed. The river may then burst its banks, causing flooding (right).

In Utah, US, in 1982-83 heavy rains caused a huge mudslide that damned Spanish Folk Canyon. A lake of water was created, which drowned the nearby town of Thistle.

LANDSLIDE CAUSES

A landslide begins when the loose material on the surface of a slope becomes unstable. Water is one of the key factors in triggering landslides. After heavy rainfall or melting snow, the surface soil becomes saturated. Water seeps through the top layer, making the layers underneath very slippery. The water increases the weight of the surface material – soil, small rocks and sediment – and weakens the forces that bind it.

As the soil loses its grip on the slope, it can no longer resist the downward pull of gravity. On steep slopes, the water-filled mass forms into a slide. On gentler slopes, it becomes a flow of earth and debris.

Erosion, where the lower part of a slope is cut or worn away, is another major cause of landslides. Erosion can result from natural factors, such as the action of the sea or rivers, or from artificial ones, such as mining and excavating. A landslide itself can be a major force of erosion, especially on steep slopes where the tree cover has been removed.

Landslides are also associated with earthquakes and volcanic activity. After the devastating earthquake in 1964 in Alaska, in the United States, much of the destruction resulted from landslides.

The river that flowed through the valley wore away the lower slopes, leaving the upper ones without support.

▼ In the Trisuli Valley in Nepal (below), the soil has been eroded from the steep mountainsides. This erosion is partly due to tree-felling operations in the area.

Water-soaked layers of shale help to lubricate the landslide.

► **A landslide has torn away the ground from beneath these houses in Antofagasta, Chile. They are left clinging to the scarp, or cliff, formed by the slide.**

VOLCANIC LANDSLIDES

On 18 May 1980, Mount St. Helens volcano in the Cascade Range of the United States erupted violently, following an earthquake beneath the volcano. As steam, hot gases and other volcanic material shot out from the volcano, the shockwaves from the earthquake triggered off two enormous landslides on the northern side of Mount St. Helens.

The falling rocks mixed with melting snow and ice, and the ash and dust from the volcano, to form a kind of mudflow called lahar. The name "lahar" originally referred to a volcanic mudflow in Indonesia, but it is now used to describe volcanic mudflows anywhere in the world.

Advance warnings had led to the evacuation of the area around Mount St. Helens. But five years later, the inhabitants of Armero and other towns and villages lying in the shadow of the Nevado del Ruiz volcano in Colombia were not so fortunate. In November 1985, a violent eruption of the volcano left around 25,000 people dead and 20,000 more without shelter.

The mudflow from the volcano eventually covered an area of more than 40 square kilometres. It destroyed the valuable farmlands of this rice- and coffee-growing area.

▼ A lahar, or volcanic mudflow, can travel over great distances at high speed. Heat from inside the volcano helps to melt the snow and ice on its slopes. The water mixes with loose material on the volcano's slopes to form a massive tongue of mud.

▲ Several weeks before the earthquake, Mount St. Helens showed signs of increasing volcanic activity. A large bulge which had appeared on the volcano's northern side was torn apart by the power of the eruption. Clouds of ash and dust poured into the atmosphere (above).

▲ Following the 1985 eruption of Nevado del Ruiz, torrents of mud raced through the streets of Armero (above) at around 160 km/hr. Cars were swept away, trees were uprooted and many of the town's inhabitants were engulfed by the tide.

DISASTER STRIKES

When a landslide begins its descent, huge quantities of soil, mud, rocks and smaller debris hurtle down a hillside. Whole villages are buried alive, and their inhabitants and livestock are swept along on a tide of mud and debris. Trees are uprooted, houses are flattened or swamped, and communications are cut off.

The mountainous regions of Nepal in Asia are prone to torrential rains. Here, the water-saturated soil has been eroded, exposing the layers of rock beneath. This erosion is worse in areas where large numbers of trees have been cut down. The hillsides are stripped of protection against the falling rain, and of tree roots to anchor the soil in place. Landslides glide over the smooth rock at speeds of up to 20 metres per second.

Following an earthquake in the Andes Mountains in Peru in 1970, a devastating mudflow swept through the Peruvian town of Yungay. Almost 18,000 people were buried alive as a giant wave of mud and debris, about 80 metres high, swamped the town.

►▲▲ Shockwaves from the 1964 Alaska earthquake caused a mudflow of clay to engulf the town.
►▲ Mud and floodwaters wreaked havoc in the streets of Armero, Colombia, after the 1985 volcanic eruption there.
► The 1970 mudflow in Yungay, Peru, caused severe damage to railway lines and roads (right).

◀ **In 1969, in the area around Los Angeles in the US, almost 112 cm of rain fell in just 42 days. These rains, some of the heaviest ever recorded in the area, washed down mud and gravel from the San Gabriel Mountains. As a swirling torrent of mud, debris and water swept over the suburb of Glendora, over 100 people were drowned and 15,000 people had to evacuate their homes. The landslide also blocked the main highway into Los Angeles.**

VAIONT, ITALY, 1963

During the night of 9 October 1963, millions of tonnes of rock slid down from Mount Toc in Italy into the Vaiont Reservoir below.

Three years previously, one of the world's highest arch dams was built below the mountain. Engineers working on the dam had noted that when the water level in the reservoir rose, a rocky mass on the mountain-side moved downwards. As the water level increased, so did the rate at which the mass moved. Heavy summer rains swelled the reservoir, and the engineers tried to halt the rate of rock movement by lowering the water level. But the rockslide could not be stopped.

At 10:41pm a torrent of rocks, mud and water raced across the wide river valley and up the opposite slope. As it fell back into the reservoir, a 100-metre high wave was catapulted over the 262-metre high wall of the Vaiont Dam into the valley below.

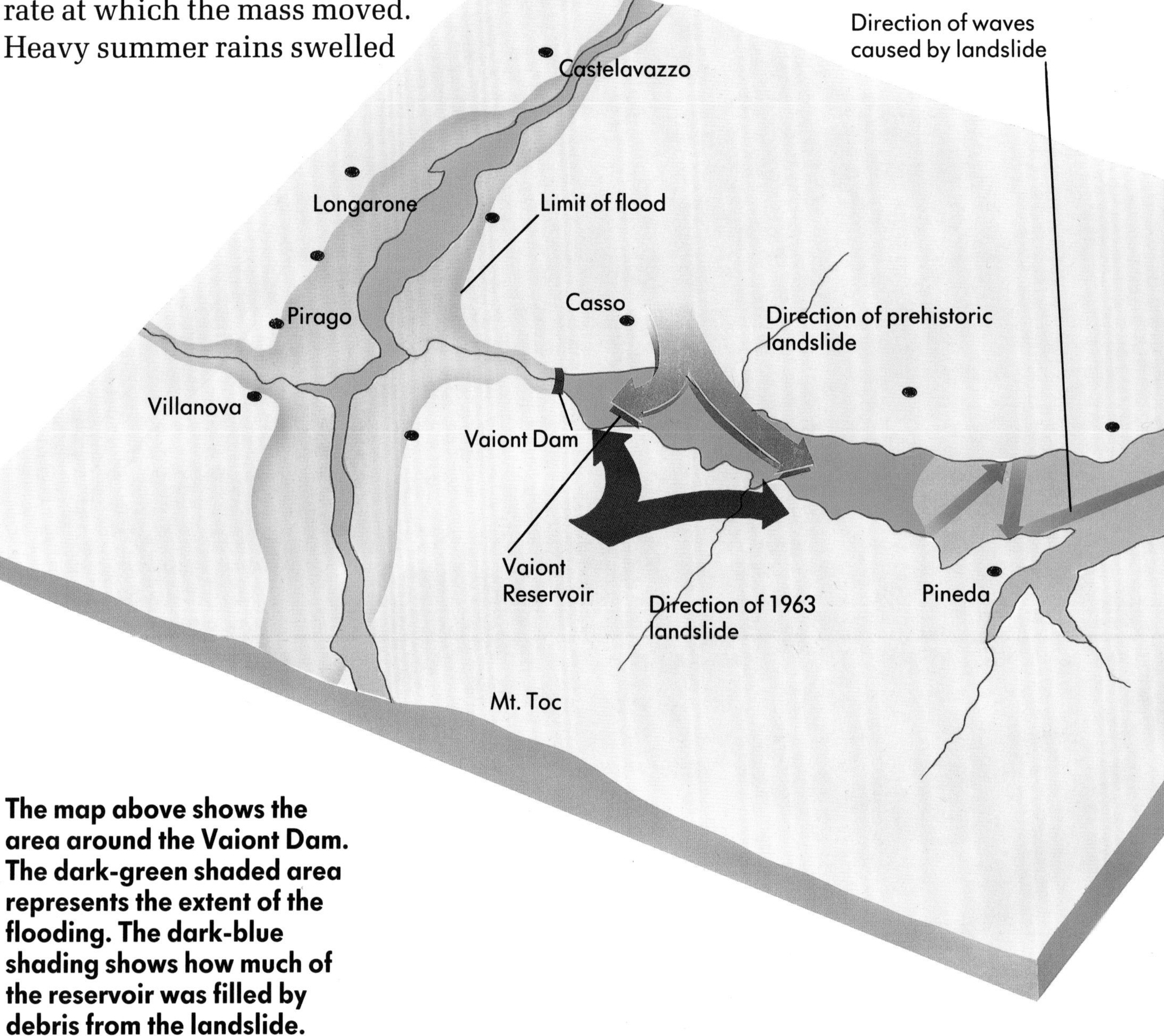

The map above shows the area around the Vaiont Dam. The dark-green shaded area represents the extent of the flooding. The dark-blue shading shows how much of the reservoir was filled by debris from the landslide.

The diagram below shows the valley of the Vaiont Reservoir before and after the landslide. In 1960, a landslide fell into the southern end of the reservoir. There is also evidence of a prehistoric landslide near Casso (see map, left).

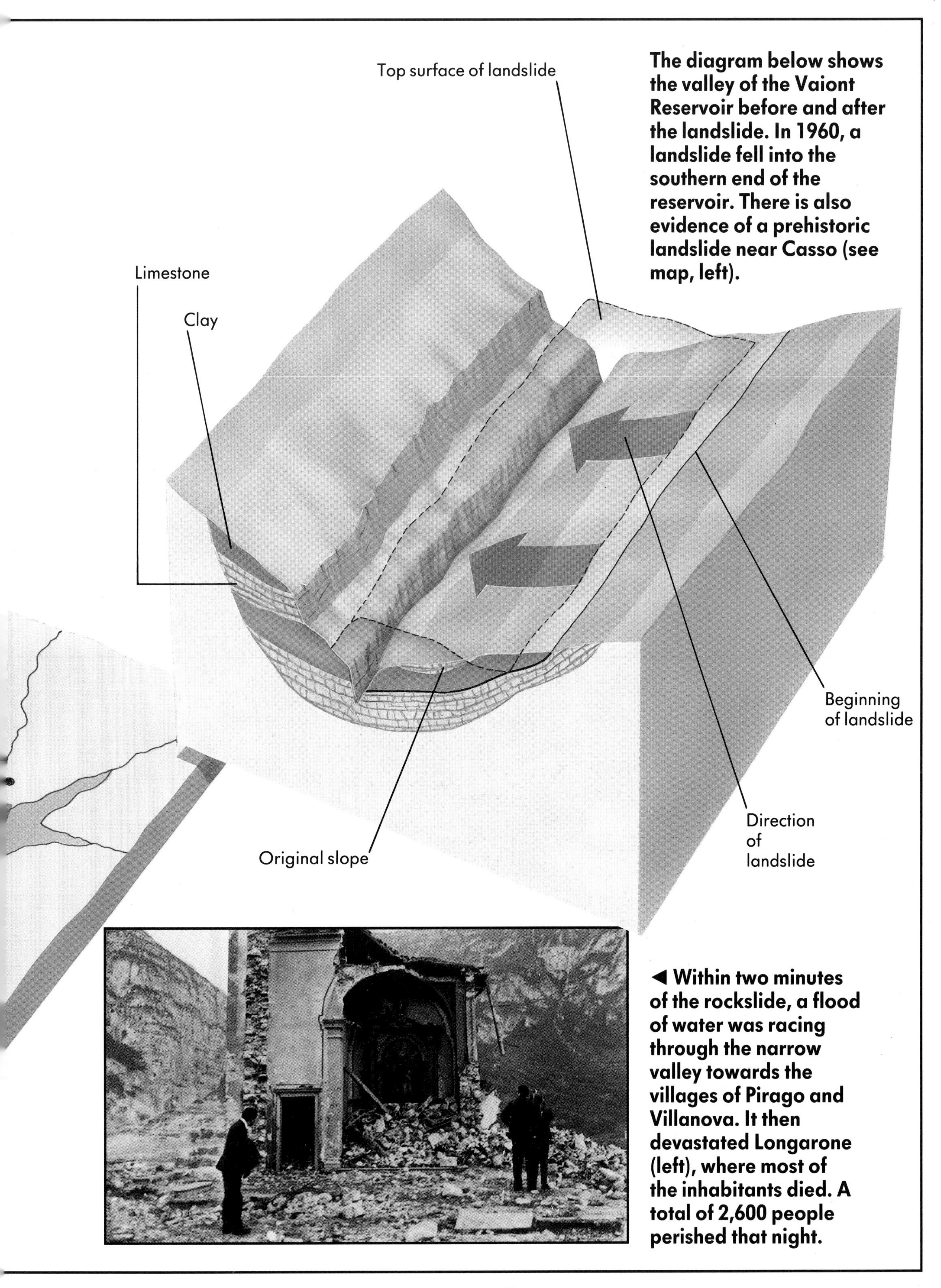

◀ Within two minutes of the rockslide, a flood of water was racing through the narrow valley towards the villages of Pirago and Villanova. It then devastated Longarone (left), where most of the inhabitants died. A total of 2,600 people perished that night.

LOOKING BACK

Evidence of avalanches and landslides dating back to prehistoric times can still be seen today. On the floor of the Saidmarreh Valley, in western Iran, lie the remains of a landslide that scientists believe occurred over 2,000 years ago. Limestone debris covers an area of around 200 square kilometres, and is 300 metres deep in places.

In 218 BC, Hannibal and his army suffered great losses from avalanches as they crossed the Alps on their way from Carthage to attack Rome. It was October, and fresh snow had fallen on a crust of old snow. Thousands of troops and animals were carried away by the avalanches triggered off as they crossed the unstable crust.

In the Middle Ages, pilgrims from northern Europe regularly crossed the Alps on their way to worship in Rome. Many believed that avalanches were acts of God, while others thought they showed the destructive power of the Devil.

Landslide!
On the morning of Christmas Day, 1839, a gaping hole opened up in the top of the cliffs at Bindon in Devon, England (left). A vast chunk of rock sank down, leaving an enormous gap about 1,200 metres long and 92 metres wide. The steep slope, or scarp, down which the landslide moved is still visible today, as are the huge lumps of displaced rock.

► During the winters of the First World War (1914-18), an estimated 60,000 soldiers fighting in the Alps were killed by avalanches. Some of these avalanches were not entirely due to natural causes. After a heavy snowfall, Austrian and Italian troops often fired into the slopes on top of the enemy below. This action released huge dry-snow avalanches. It was the first time in military history that avalanches had been used as a weapon.

◄ Hannibal's treacherous journey across the Alps involved about 38,000 foot soldiers, 8,000 horses and over 30 elephants. According to the Roman historian Livius, more than 18,000 men and 2,000 horses were lost during the crossing, mainly because of avalanches. The reduced numbers of Hannibal's troops contributed to their eventual defeat by the Roman army.

ARE WE CAUSING MORE?

We cannot prevent natural disasters such as avalanches and landslides. Yet we can stop or reduce the human activities that make it more likely that one of these powerful forces will strike. When large numbers of trees are cut down, to provide fuel and timber, as well as land for farming, nothing remains to trap the rainwater and hold the surface of the slopes in place. This deforestation also removes the natural barrier to avalanches and landslides that trees and other vegetation provide.

Mountainous areas in other parts of the world are also being cleared and developed as tourist resorts. Millions of people then visit the mountains to sightsee and explore. These activities can wear away the soil and vegetation, making the area more prone to avalanches and landslides.

Road construction
Excavation work for new road building steepens slopes and removes support for the upper layers.

Quarrying and mining
As large amounts of natural resources, such as stone and coal, are removed from the ground, the surrounding land becomes unstable and may collapse.

Deforestation
The roots of trees and ground foliage hold water like a sponge, releasing it slowly into the surrounding soil in a controlled flow. When a hillside is stripped of its cover, the exposed soil erodes very quickly.

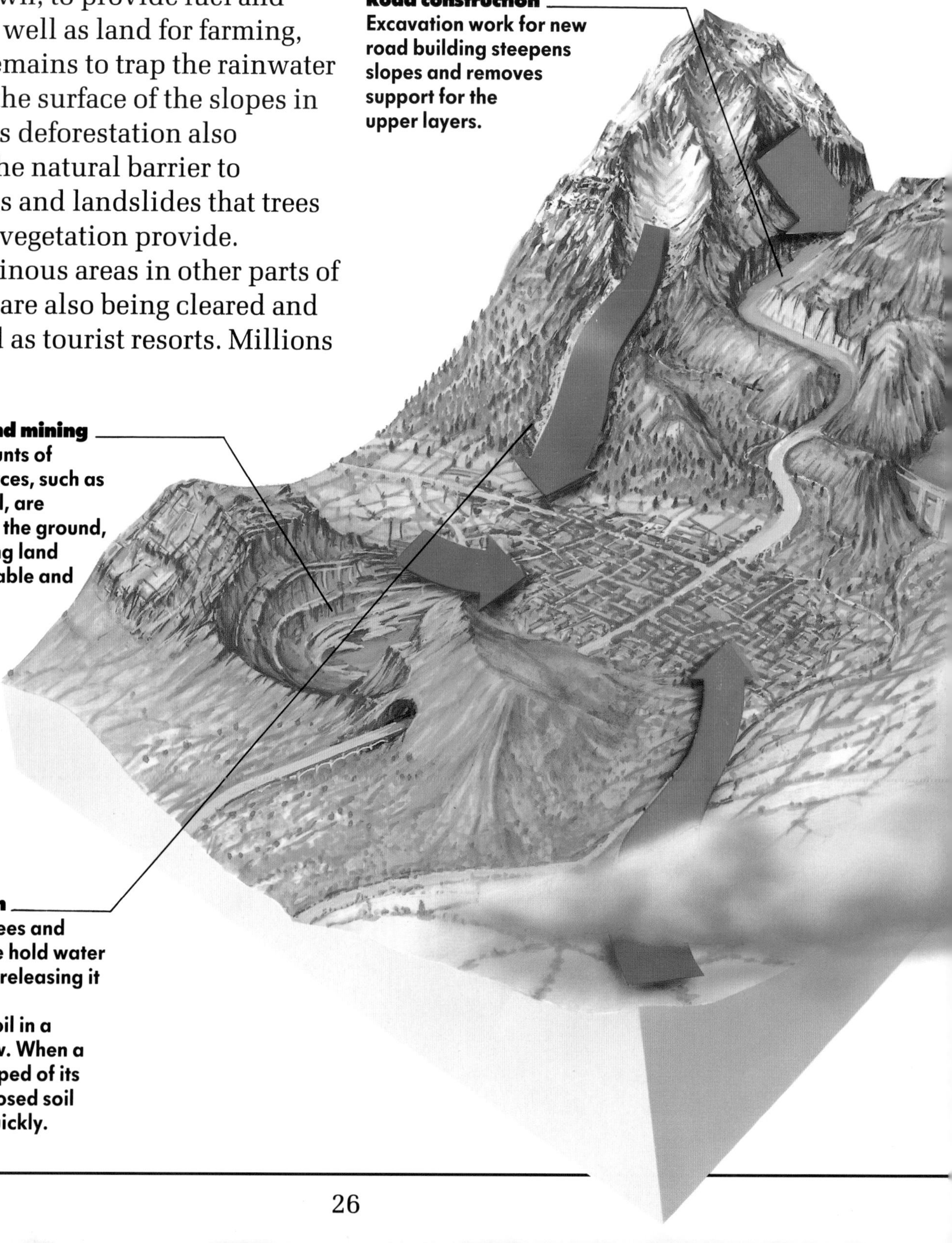

The threat of landslides and avalanches is increased by excavation and construction work, and by the removal of natural resources from the ground. The shape of a slope may be steepened or altered to such an extent that it becomes unstable. Heavy rains and snow add further weight to the slope, weakening the surface soil's grip on the hillside.

▲ Trees help to prevent the erosion of soil by heavy rains. Deforestation has caused this landslide in Malaysia (above). The slide has completely destroyed a main road.

Reservoirs
The construction of reservoirs in narrow mountain valleys that are prone to landslides increases the risk of a flood disaster.

Global warming
As the atmosphere traps increasing amounts of the greenhouse gases that warm the Earth, changes in weather patterns may occur. Very heavy rains could increase the number of mudflows and landslides.

Forest fires
A lack of hillside vegetation is blamed for the mudflows that are common in parts of southern California in the United States. Forest fires destroy the trees and shrubs on the steep slopes around many residential areas. After heavy rains, the topsoil becomes saturated with water and starts to slip downhill.

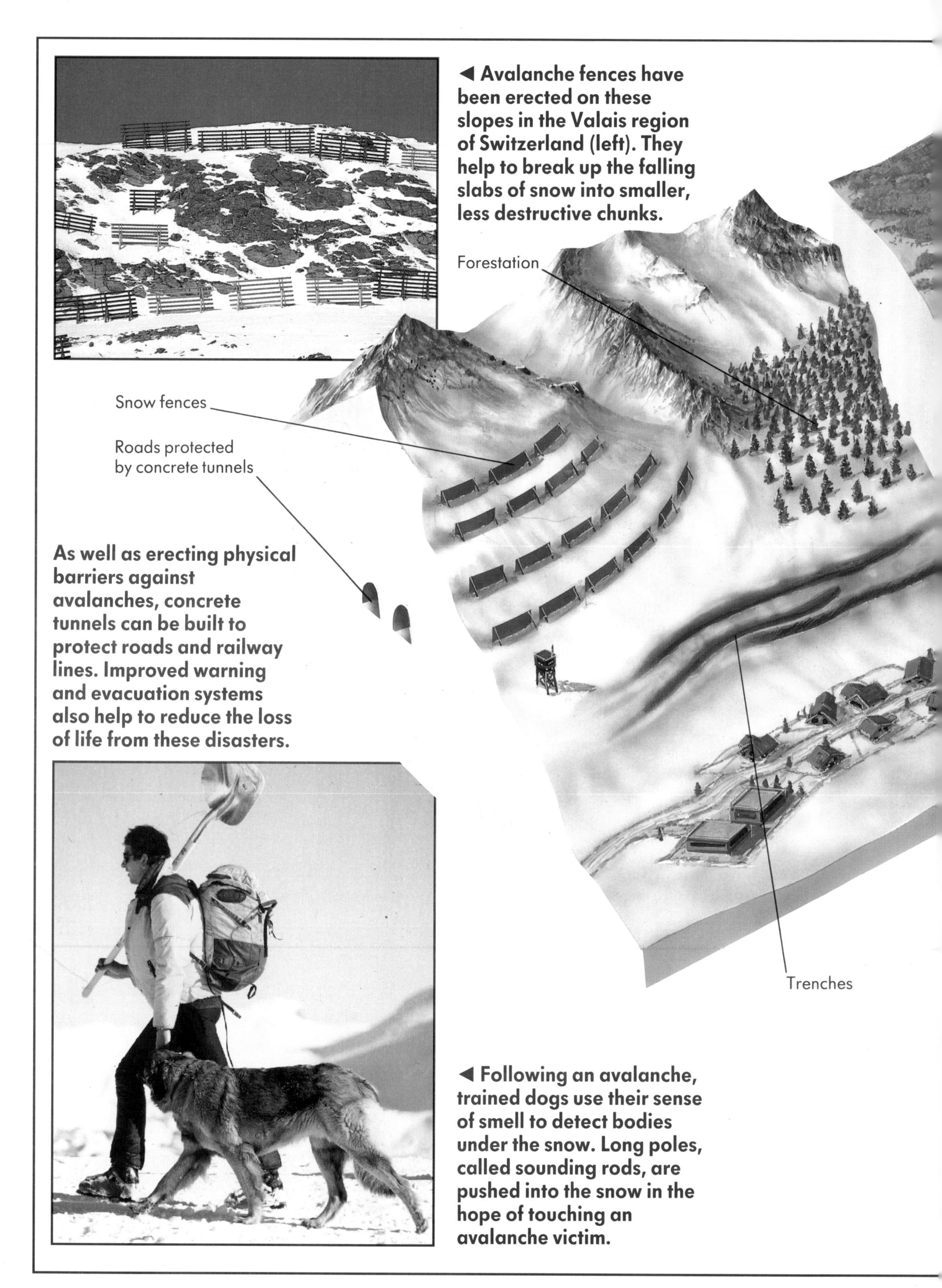

◀ Avalanche fences have been erected on these slopes in the Valais region of Switzerland (left). They help to break up the falling slabs of snow into smaller, less destructive chunks.

As well as erecting physical barriers against avalanches, concrete tunnels can be built to protect roads and railway lines. Improved warning and evacuation systems also help to reduce the loss of life from these disasters.

◀ Following an avalanche, trained dogs use their sense of smell to detect bodies under the snow. Long poles, called sounding rods, are pushed into the snow in the hope of touching an avalanche victim.

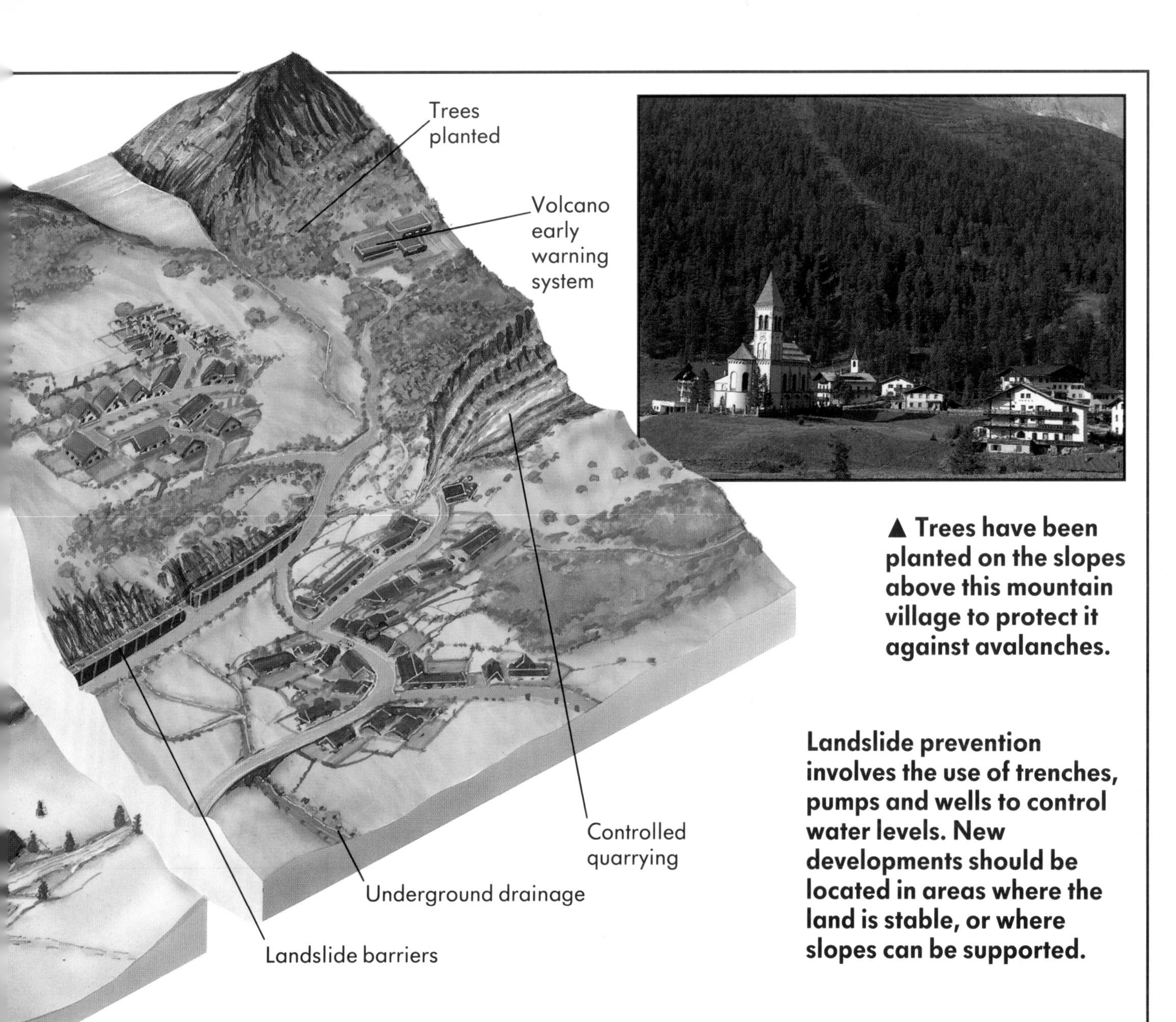

▲ Trees have been planted on the slopes above this mountain village to protect it against avalanches.

Landslide prevention involves the use of trenches, pumps and wells to control water levels. New developments should be located in areas where the land is stable, or where slopes can be supported.

WHAT CAN WE DO?

Many different measures can be taken to reduce the devastating impact of an avalanche or landslide. In Switzerland, controls include setting off explosives to release avalanches artificially, and planting triangular patches of trees above mountain villages. Anti-avalanche constructions, such as trees, fences or barriers, are used to slow down, deflect or break up the falling snow.

Water is one of the prime causes of landslides. Measures can be taken to control and monitor the amount and pressure of the water in unstable slopes. Underground drainage is also installed in some places to reduce the water content of landslide-prone hillsides. The lower end, or toe, of a slide can be shored up with concrete, rock or soil.

In the United States, scientists are using computers to try and forecast avalanches. The study of volcanoes can also help to predict volcanic eruptions more accurately, thus reducing the death toll from volcanic mudflows.

FACT FILE

The world's largest landslide
The largest landslide in the world in the past 2 million years is believed to have occurred in the United States. At the foot of Mount Shasta, an active volcano in the state of California, lies an area of debris covering some 450 square kilometres. The debris is believed to date from a rockfall that happened between 300,000 and 360,000 years ago.

Recent avalanches and landslides
1990
Peru – following heavy rains, a mudslide buried the village of San Miguel de Río Mayo in the jungle region 800 km north of the capital, Lima. A total of 200 people were reported missing.

1991
Malawi – after flooding and a series of mudslides in the south-east of the country in March, 516 were reported dead or missing, and between 40,000 and 50,000 people were left homeless.
New Zealand – a huge avalanche on 14 December almost destroyed the eastern face of Mount Cook, the highest mountain in New Zealand. Thousands of tonnes of ice, snow and rock hurtled down from the summit of the 3,764-metre high mountain, and smashed into the Tasman Glacier, a popular tourist spot, over 6 km below.

1992
Argentina – torrential rain in January caused a debris avalanche of rocks, mud and trees to destroy the town of San Carlos Minas. Sixty people were believed to have been killed.
Turkey – avalanches in February in the remote Kurdish Mountains of south-eastern Turkey left more than 150 people dead. In the village of Gormec, 71 soldiers died under a single avalanche. Rescue workers were delayed by blocked roads and strong blizzards.
Israel – tonnes of rocks and soil fell on a cafe in East Jerusalem on 29 February, leaving 23 people dead and a further 20 injured. After heavy snow and rain, part of the hillside above the cafe collapsed and burst through a wall onto the cafe below.

Avalanche control in Canada
In Canada's Selkirk Mountains, one of the world's largest anti-avalanche programmes is operated along Rogers Pass, with the help of the Royal Canadian Horse Artillery. In a densely populated mountain area, the soldiers use gunfire to release avalanches and protect a 40-km stretch of major road from 160 separate avalanche paths.

Mudflow technology
Japan has become a world leader in the study and prevention of mudflows. Since 10 per cent of the world's active volcanoes are found in Japan, many of the country's population centres are threatened by mudflows. On the side of Mount Usu on the island of Hokkaido, steel and concrete structures have been erected to trap falling rocks and slow down the flow of mud and other debris. Elsewhere, television monitors and measuring equipment help to detect volcanic activity and the onset of a mudflow or other landslide.

Volcanic mudflows
On the slopes of Mount Rainier, a volcano in the United States, more than 55 mudflows have been recorded in the past 10,000 years.

Landslide deaths
Between 25 and 50 people die in the United States each year from landslides.

Mining waste disaster
The waste material from mining and excavating can be a serious landslide hazard. For many years, the waste from the coal mines around Aberfan in South Wales was piled dangerously high. When it became saturated with water following heavy rains in October 1966, part of the tip collapsed. The waste material slid downhill, killing over 144 people including 116 children in the local school.

GLOSSARY

avalanche – the downward movement of a mass of ice and snow.

creep – the slow, almost unnoticeable, movement of loose soil down the surface of a slope.

debris – loose solid material such as soil, mud, small pieces of rock and stones.

deforestation – the large-scale removal of trees and other vegetation.

dry-snow avalanche – a huge blast of white powdery snow that travels at great speed along or above the ground.

earthquake – a violent shaking movement of the outer layer, or crust, that surrounds the Earth.

earth tremor – the violent trembling of the ground that occurs during and after an earthquake.

erosion – the removal or gradual destruction of a surface by the action of water, ice or wind. Human activities, such as tree-felling, can also lead to erosion.

eruption – the violent release of steam, rocks, dust and ash through an opening in the Earth's surface.

global warming – an increase in the Earth's temperature due to a build-up of greenhouse gases, such as carbon dioxide, in the atmosphere.

greenhouse gas – one of the gases in the atmosphere that traps heat from the Sun and keeps the Earth warm. Greenhouse gases include carbon dioxide and water vapour.

groundwater – water that is found in rocks and soil beneath the surface.

lahar – a mixture of rocks, melting snow and ice, and the ash and dust from an erupting volcano.

landslide – the downhill movement of large amounts of rock and soil.

lubricate – to make a surface slippery.

mudflow – the rapid downhill movement of soil mixed with water.

rockfall – the downward movement, usually through the air, of rocks and other material from a cliff or steep slope.

rockslide – the slipping movement of rock along one or more surfaces.

sandstone – a coarse rock made up of grains of sand joined together.

saturated – filled with the maximum amount of a liquid or solid substance.

scarp – the steep slope left behind when a landslide falls away.

shale – a kind of soft rock made from tiny pieces of mud and clay.

shock wave – a burst of energy that is released from the centre of an earthquake.

slab avalanche – the downward movement of a large chunk of solid snow.

slump – the movement down a slope of a block of rock and soil. The slumped block tilts back towards the slope.

unstable – describes a slope that is likely to start moving downwards.

volcano – an opening in the Earth's surface through which hot rocks and gases can escape.

wet-snow avalanche – the downhill sliding movement of large chunks of wet snow.

INDEX

Photographic credits:
Cover and pages 10, 19 top left, 20 top and 27 top: Frank Lane Picture Agency; title page and page 20 bottom: Robert Harding Photo Library; pages 4-5 and 9: Science Photo Library; page 13 top: Paul Galvin; page 13 bottom: Charles de Vere; pages 15, 17, 19 top right and 20 middle: Frank Spooner Pictures; pages 16 and 28 top: J. Allan Cash Photo Library; page 18-19: Natural History Photographic Agency; page 23: Topham Picture Source; pages 24 and 25 bottom: Mary Evans Picture Library; page 25 top: Roger Vlitos; pages 27 bottom and 28 bottom: Spectrum Colour Library; page 29: Geoscience Features.

PRINTED IN BELGIUM BY proost INTERNATIONAL BOOK PRODUCTION